Prehistoric Mammals
of Western Australia

Ken McNamara and Peter Murray

T0358114

GOVERNMENT OF
WESTERN AUSTRALIA

WESTERN AUSTRALIAN

mUSEum

Contents

Left: The usual way of entering a cave on the Nullarbor Plain
(photo C. Bryce).

Introduction

During the construction of a path in Mammoth Cave, south of Margaret River, in 1904, a number of bones were unearthed. Immediately recognisable among the bones were fragments from an extinct form of large kangaroo, previously unknown in Western Australia. Before that time, evidence for the presence of extinct mammals in Western Australia was based on a few bones of the extinct giant marsupial *Diprotodon* that were found in the Kimberley, Balladonia and near Lake Darlot, north of Leonora.

It was not until 1909 that Mr Ludwig Glauert, keeper of geology at the Western Australian Museum, was able to excavate the bone-bearing deposits in Mammoth Cave. What he found over six years of excavations was an exceedingly rich accumulation of many thousands of bones, representing many species of mammals. Some were the remains

Left: Entrance to Mammoth Cave *(photo K. McNamara)*.

of small species still living today, but many of the bones far exceeded in size those of any modern-day native Australian mammal. This find supported the evidence which had been accumulating during the latter part of the nineteenth century in eastern Australia; namely, that in prehistoric times giant mammals had roamed the Australian bush, but had become extinct. A treasure trove of bones had been discovered: a time capsule containing a mammal fauna of long ago.

After examining the bones back in Perth, Glauert soon realised that the diversity of mammalian forms that once lived in the extreme south-west of Australia many thousands of years ago far exceeded that of today. There were bones of a marsupial the size of a buffalo, large kangaroos more than two metres tall, wallabies much bigger than any living species, a marsupial 'lion' about the size of a leopard, giant echidnas and wombats; plus the thylacine (Tasmanian tiger), Tasmanian devil, koala and other smaller marsupials, some of which are still living in the south-west today.

Right: A complete skeleton of *Thylacoleo* in a Nullarbor cave *(photo C. Bryce)*.

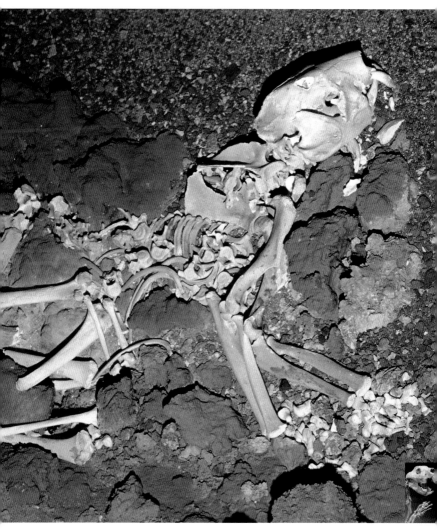

These large, extinct animals are known as the 'megafauna'. These are usually defined as animals that weighed more than 40 kilograms, or were more than 30 percent larger than living relatives.

> And from their broken bones and teeth, what can we deduce about the way they lived?

As well as megafaunal mammals, there were also megafaunal reptiles and birds: the huge 6–7-metre long varanid lizard *Megalania* that weighed over half a tonne; the 5-metre long python *Wonambi*; and large, flightless birds such as *Genyornis*, which weighed more than 100 kilograms.

The discovery and subsequent excavation of

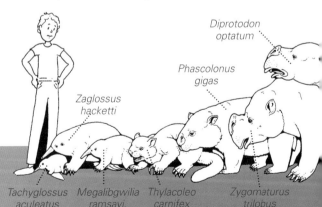

Diprotodon optatum

Phascolonus gigas

Zaglossus hacketti

Zygomaturus trilobus

Tachyglossus aculeatus   *Megalibgwilia ramsayi.*   *Thylacoleo carnifex*

about 30 cubic metres of bone-bearing deposits in Mammoth Cave between 1909 and 1915 was to be the first of many such megafaunal finds made in south-western Australia. The Mammoth Cave deposit and several subsequent discoveries in caves in the Nullarbor region have allowed scientists to reconstruct an ancient Western Australian fauna that lived tens of thousands of years ago and was quite different from the fauna of today. Australia must have looked a vastly different place to the first humans when they arrived on the continent.

What did these extinct animals look like? And from their broken bones and teeth, what can we deduce about the way they lived? This book aims to provide a basic guide to the major types of larger extinct mammals that once inhabited Western

Protemnodon

Left: Illustration by Jill Ruse.

Simosthenurus   *Thylacinus cynocephalus*

Australia. It describes the major features of these animals and presents modern ideas on how they might have lived, what they might have fed on and how they might have interacted with one another. A final consideration, perhaps the most perplexing question of all, is why did so many animals, particularly the larger types, become extinct in a relatively short period of time?

As with all reconstructions of extinct animals, it is necessary to consider the question of how the animal remains are preserved; which parts of the animal are preserved; which parts of the animals reveal most to us about their life habits; and under what conditions, and in what environments, preservation of their remains took place.

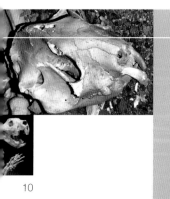

factfocus

Our understanding of the extinct large mammals that once inhabited Western Australia is mostly derived from their skeletal remains. Rarely, as with some specimens from the Nullarbor Plain, entire specimens may be preserved, complete with fur, tongue, eyeballs and whiskers.

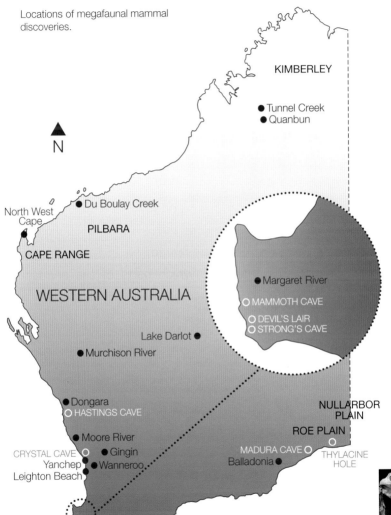

Locations of megafaunal mammal discoveries.

N

KIMBERLEY

● Tunnel Creek
● Quanbun

● Du Boulay Creek

North West Cape

PILBARA

CAPE RANGE

WESTERN AUSTRALIA

● Margaret River
○ MAMMOTH CAVE
○ DEVIL'S LAIR
○ STRONG'S CAVE

Lake Darlot ●

● Murchison River

● Dongara
○ HASTINGS CAVE

NULLARBOR PLAIN

ROE PLAIN

● Moore River

CRYSTAL CAVE ○ ● Gingin
Yanchep ● ● Wanneroo
Leighton Beach

MADURA CAVE ○
Balladonia ●

○ THYLACINE HOLE

The Fossil Remains:
Their Occurrence and Preservation

The word 'fossil' literally means an object that has been dug up, so strictly speaking any bone dug out of the ground is a fossil. If the bone is in the ground for a sufficiently long period of time, mineral-laden waters may percolate through the porous boney material, saturating it with mineral deposits. It is then 'petrified' and will become much heavier and more robust. The quicker this process of petrification, the better chance the bone has of surviving. Petrification does not occur where the groundwater is acidic. Under certain conditions bone can survive being buried for long periods of time without becoming mineralised, but only if it is protected from the corrosive action of acidic groundwater. Many of the bones found in Western Australian caves are preserved like this.

Left: Careful excavation of a Nullarbor *Thylacoleo* in the hope of finding fossilised DNA *(photo C. Bryce)*.

Certain kinds of bones will be preserved more readily than others. Delicate skull bones are readily crushed and are less commonly preserved, particularly in older deposits. The more durable jawbones, teeth and limb bones often persist. This is fortunate, as most mammals can be identified by the form of their dentition. Molars, premolars, canines and incisors differ greatly in number, size and shape from one species to another. Certain teeth may be greatly enlarged or have characteristic shapes in some mammals to serve a particular function. In addition to providing invaluable information on the identity of the animal, the nature of the dentition is of great value in indicating an extinct animal's eating habits, and hence its way of life. Very often a single tooth, or even just a fragment of a tooth, is sufficient to allow palaeontologists to identify the species and help build up a picture of animal life within a particular region. Whereas many of the fossilised remains of mammals found in Western Australia are isolated bones, in the last fifteen years complete skeletons of some of these extinct giants have been found in old river deposits and caves.

Above: Ludwig Glauert, Curator of Geology, in the Western Australian Museum not long after collecting bones from Mammoth Cave.

The chance of most land-dwelling animals becoming fossilised is remote. Once an animal dies, predators and scavengers quickly move in on the corpse and clean the flesh from the bones, which become broken and scattered. Once exposed to the sun and rain, the bone rapidly breaks down, and the remains of even large animals disappear within a year or so. However, there are a number of circumstances whereby bone can be preserved and become fossilised. Most of these situations have one factor in common: they occur in areas where the bone is rapidly covered by sediment, or otherwise protected from the harmful effects of rain and sunlight.

Caves

The majority of fossil bones in Western Australia have come, like the Mammoth Cave bones, from floor deposits of limestone caves. The calcium carbonate in the cave soil helps preserve bone. At least eighty caves in south-western Australia, and more than thirty caves on the southern part of the Nullarbor Plain, have yielded mammal remains, as have a few in the Cape Range and the Kimberley.

Bones have accumulated in these caves through a number of different processes. They may have been washed in by streams, then been rapidly covered by sediment during floods. Other cave material may have been derived from animals that fell into the cave through openings leading to the surface. Frequently, openings into large, subterranean chambers may take the shape of steep or vertical narrow pipes. It is possible that many animals, as they wandered, or jumped, through the bush, did not see the openings and so tumbled down the pipes to their death. Accumulations of bone-bearing deposits like those

Right: Dr Gavin Prideaux carefully collecting small bones in a Nullarbor cave *(photo C. Bryce)*.

found in Mammoth Cave, where there are a large proportion of bones of the large, extinct species, may have formed this way. An alternative suggestion for the bones in Mammoth Cave is that they were brought there by early humans; indentations, breakage patterns and charring of some bones seem to have been man-made. It is possible that a small proportion of the bones in Mammoth Cave were left there by humans who, after having hunted and killed their prey, had then cooked and eaten it in the cave.

… the mummification and exquisite preservation of a number of animals …

Preservation of animals in caves can also occur by the much rarer process of mummification. If the

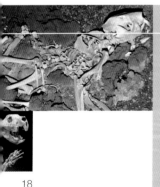

factf(oc)us

Much of what we know about the extinct megafauna of Western Australia comes from bones collected from caves. Many have been found just lying on the cave floors, but by digging down through layers of accumulated sediment it is possible to see how the fauna has evolved over time, as environments and climate have changed.

environmental conditions in a cave are at a low level of humidity, a relatively low temperature, and also perhaps a high salinity, then entire animals — flesh, fur and hair — may be preserved. Such conditions prevailed for over 4,000 years in one cave on the Nullarbor, and resulted in the mummification and exquisite preservation of a number of animals, now vanished from the region, including a thylacine.

Rivers

Preservation of bones has also occurred in environments other than caves. Some of the best preserved fossil bones in Western Australia have come from river deposits — silts, sand and gravel — laid down by rivers in earlier times and now preserved as raised terraces above the present river level. Large animals caught in a flooded river, heavy with sediment, would soon have been covered by the sediment and so protected from destruction. Thousands of years later, erosion into the old river terrace, perhaps by the river changing its course or by the activity of humans, could have led to the uncovering of the remains of animals long since vanished from the face of the Earth.

Lakes

Some lake deposits may also yield fossilised bones. Although not many bones have been recovered from lake sediments in Western Australia, many remains of extinct, large marsupials have been excavated from lake sediments in inland areas of eastern Australia, such as Lake Callabonna. Many of these are now salt lakes. Large animals walking on seemingly hard surfaces around the lake would have broken through the veneer of hard, salt-caked surface material and become bogged in the great thickness of soft mud beneath. Unable to free themselves, they would soon have perished. Excavations have revealed a large number of complete skeletons of extinct, giant marsupials preserved in this manner.

Excavations have revealed a large number of complete skeletons …

In Western Australia, water soaks in the granite country around Balladonia attract many animals today, as they have done for thousands of years. Numerous animals have perished at these sites, perhaps during periods of drought, and large quantities of bones have accumulated over long periods. These deposits provide a rich source of bone material for the study of ancient animals.

Dunes

The sandy limestones near the coast of Western Australia have also yielded occasional fossilised bones. This limestone represents, in part, a series of old sand dunes (see another book in this series entitled *Pinnacles*). During periods of dune standstill, soil horizons and vegetation are able to develop before being inundated at a later date by another mobile dune. Within these fossil soils, bones and teeth have occasionally been found.

factf(ocu)s

The remains of the extinct megafauna of Western Australia are found not only in caves, but also in river terrace sediments, fossil soil horizons and in sediments deposited in lakes or in water soaks around igneous rock outcrops. All of these remains contribute to our understanding of this vanished fauna.

Age of the Western Australian Fossil Remains

Dating the deposits which contain the fossil bones is often very difficult. By referring to deposits of known age in other parts of Australia, it is possible, on the basis of the fossils within those deposits, to be reasonably certain that nearly all sites in Western Australia that contain the extinct large mammals are Pleistocene in age. That makes them between 1.81 million and 10,000 years old. On the basis of purely faunal evidence, many of the Western Australian fossils are Late Pleistocene in age — in other words, 126,000 years or less — but some are older. Recent finds from the Nullarbor contain different species and are probably much older.

A recent analysis of the age of megafaunal remains from caves in south-west Australia has

Left: (*photo ©iStockphoto.com/EdeWolf*).

shown that they range in age between 46,000 and >212,000 years. Specimens from Devil's Lair in the Margaret River region have been dated at 47–48,000 years old. Those from Mammoth Cave have been dated at 55–74,000 years old. These dates were obtained using methods that analyse the amount of decay of uranium into thorium, and optical dating. This latter method is based on the fact that electrons in buried grains of quartz become excited to higher energy levels over time, due to their exposure to radioactive elements in the surrounding sediments. By measuring the exposure time, it is possible to predict when the grains, and therefore the associated bones, were buried.

Fossil bones from a deposit at Quanbun in the

 factf⦶cus

Most of the fossil bones found in caves in south-western Australia date between 46,000 and more than 212,000 years old. The oldest remains in the state, found in the Kimberley, may be as much as 5 million years old. Caves on the Nullarbor contain fossil bones more than 400,000 years old.

Above: Entrance to a cave on the Nullarbor Plain *(photo C. Bryce)*.

Kimberley indicate the existence of deposits older than those occurring in the south of the state. This northern deposit is perhaps Early–Middle Pliocene in age (5–2 million years old). This age is based mainly on the presence of one species of extinct large kangaroo, *Macropus pan*, which in other parts of Australia is only known from deposits of Pliocene age.

In 2002 a spectacular menagerie of fossil bones was discovered in caves in the eastern Nullarbor. Sixty-nine species of vertebrates have been identified to date, some of which are megafauna. Dating of the sediments in the caves suggests an age in excess of 400,000 years.

Following Page: Reconstruction of the Pleistocene megafauna in Australia by Jill Ruse.

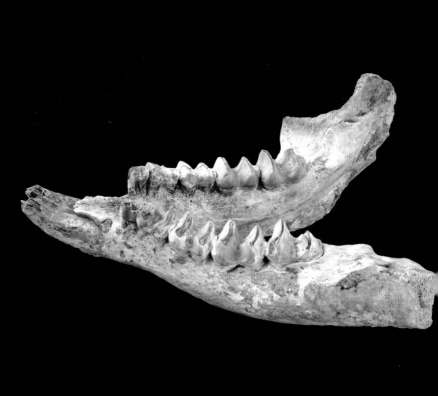

Diprotodontids

Diprotodon

Before the early 1990s, a small number of bone fragments, teeth and lower jaws from areas as far apart as the Kimberley, the Pilbara, Lake Darlot north of Leonora, and Balladonia at the western edge of the Nullarbor Plain were sufficient to tell us that the largest marsupial that ever walked the Earth once lived in Western Australia. Its name is *Diprotodon optatum*. The name *Diprotodon* means 'two forward teeth' and alludes to the incisors of the lower jaw, which point straight forward. But it wasn't until 1991 that a complete skeleton was found in Du Boulay Creek, near the mouth of the Fortescue River in the Pilbara. Measuring about three metres long and standing more than two metres high, this particular individual is at least 80,000 years old and one of the largest *Diprotodon* specimens (and therefore one of

Left: Lower jaw of *Zygomaturus trilobus*, Murchison River; one third natural size *(photo G. Deacon)*.

the largest marsupials) ever found.

Diprotodon was one of the last surviving members of a family of large herbivorous (plant-eating) marsupials, the Diprotodontidae, which first appeared more than 24 million years ago. These first diprotodontids were only about the size of sheep, but the Late Pleistocene *Diprotodon optatum*, measuring nearly two metres high at the shoulder and 2.5 metres in length, and weighing about two tonnes, was roughly the size of a rhinoceros. It was a heavily built animal, with stout limbs and a moderately long neck. It is believed to have looked rather like an oversized, long-legged wombat. Indeed, wombats are the nearest living relatives of *Diprotodon*, although their relationship is very distant. Several skull characters are shared by *Diprotodon* and wombats, which reflect the evolution of the two forms from a common ancestor that lived more than 30 million years ago.

Right: Reconstruction of *Diprotodon optatum* by Jill Ruse.

Diprotodon was probably quite a slow-moving animal. One of its peculiarities was the nature of its feet. At the end of its stout limbs and massive wrists and ankles were absurdly small toes. The weight of the large body must have been borne mainly by the bones of the wrist and ankle. In the hind feet the big toe opposed the others.

The brain of *Diprotodon*, although larger than the brain of its ancestors, was smaller relative to the whole body size. However, the brain was housed in a very large head. Much of this large skull was not brain case but an open network of air sinuses that acted to lessen the weight of the large skull. The brain itself was set within an inner brain case inside the massive skull. This skull structure was important because it allowed the external shape of the skull to remain much the same as its much smaller ancestors, and was necessary to accommodate the huge teeth and muscles needed to process large amounts of coarse vegetation. Much of the bone of the skull was very thin, further reducing the weight.

The development of the very large skull probably

fact focus

Diprotodon is the largest known marsupial, reaching three metres in length and standing about two metres tall. A herbivore, *Diprotodon* is likely to have lived in large herds. Its closest living relative is the wombat. The structure of its feet suggest it was a slow mover and so prone to attack by predators.

Above: Molar teeth of *Diprotodon* (left) and *Zygomaturus* (right); natural size *(photo D. Elford)*.

relates to an increase in the relative size of the teeth required to process enough food to support such a large body mass. It would also have adequately withstood stresses from chewing very coarse vegetation. The larger skull would have been supported by relatively larger muscles than were possessed by smaller ancestral diprotodontids.

In addition to its pair of lower, elongate incisors, there was a pair of chisel-like incisors in the upper jaw of *Diprotodon*, which opposed the lower pair. The molars, which reached up to five centimetres in length, were characterised by their high and narrow transverse double-ridged structure. Opposing molars met with a vertical slicing and grinding motion, ideally suited to shredding the tough vegetation on which *Diprotodon* fed exclusively. This probably consisted of leaves and shoots of small plants and trees.

There is some debate about the nature of the snout region of *Diprotodon*. An unusual vertical plate of bone, not found in any living mammal, has posed some problems when attempts have been made to reconstruct the animal. It has been suggested that this plate supported muscles which operated either very large, mobile lips, or even a trunk, perhaps giving the animal a tapir-like appearance. Impressions of the skin, hair and footpads are known but no evidence of the nature of the nose or mouth has been found. The combination of very flexible lips and long, narrow lower incisors

… one of the last surviving members of a family of large herbivorous marsupials …

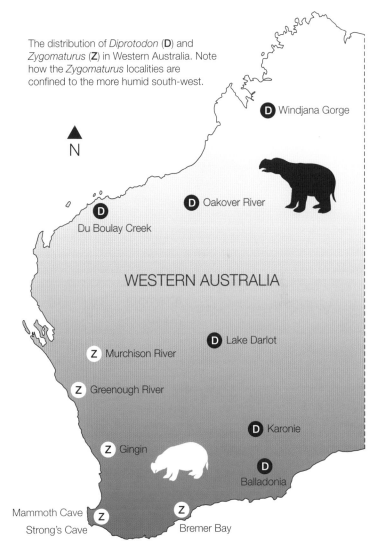

The distribution of *Diprotodon* (**D**) and *Zygomaturus* (**Z**) in Western Australia. Note how the *Zygomaturus* localities are confined to the more humid south-west.

N

D Windjana Gorge

D Oakover River

D Du Boulay Creek

WESTERN AUSTRALIA

Z Murchison River

D Lake Darlot

Z Greenough River

D Karonie

Z Gingin

D Balladonia

Mammoth Cave
Strong's Cave
Z

Z Bremer Bay

would have allowed leaves to be easily stripped from shrubs and trees.

It has been suggested that the large size attained by forms such as *Diprotodon* occurred as a response to poorer food sources caused by the climate becoming drier. Certainly, during Late Cenozoic times, 15 million to two million years ago, the climate over much of Australia became progressively more arid. This favoured the spread of *Diprotodon* as the forest areas contracted. Whether or not increasing aridity during the last two million years led to the eventual extinction of *Diprotodon* and other large mammals is open to debate (see page 102).

Many complete skeletons of *Diprotodon* have been found in Lake Callabonna in South Australia. Following the discovery of bones in 1892, parts of 360 individual skeletons were recovered in a limited area of only a few hectares. These animals had become mired in the sticky clays around the lake's edge. This large assemblage of individuals tells us a number of things about the animal, apart from what its skeleton looked like. Firstly, *Diprotodon* was a relatively common animal and was probably

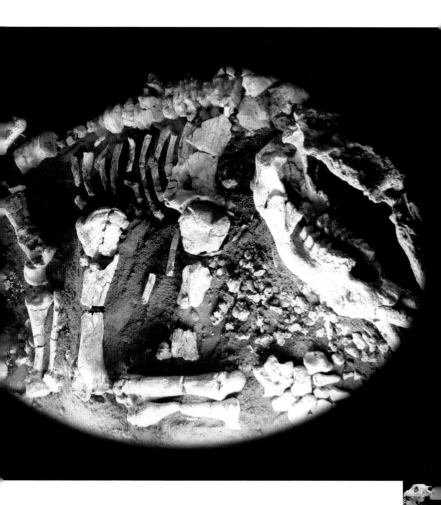

Above: Skeleton of *Diprotodon* from Du Boulay Creek in the Pilbara on display at the Western Australian Museum *(photo G. Deacon)*.

Above: Remains of *Diprotodon* in Du Boulay Creek *(photo J. Long)*.

a social creature and moved around in small herds. Secondly, the female *Diprotodon* possessed a pouch; remains of juveniles have been found just in front of the pelvis of adult specimens. Thirdly, vegetation found in the rib cage of some specimens, in the position where the stomach would have been, gives a vivid illustration of the last meal of some of these giant marsupials. They fed on saltbush and other shrubs, which would have been growing around the edges of the lake.

Below: Reconstruction of *Zygomaturus trilobus* by Jill Ruse.

Zygomaturus

Whereas *Diprotodon* occupied the arid areas of Western Australia, another large diprotodontid which lived in Western Australia was restricted to the semi-arid and humid coastal regions of the south-west. This animal, hundreds of whose bones have been collected from Mammoth Cave, is called *Zygomaturus trilobus*.

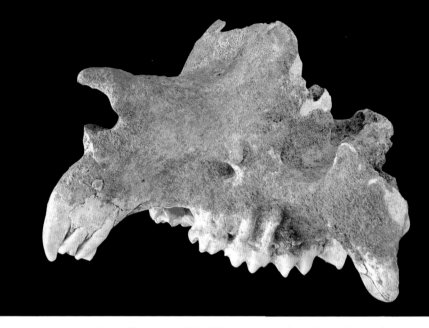

Above: Upper mandible of *Zygomaturus trilobus* from Mammoth Cave; one third natural size *(photo G. Deacon)*.

Zygomaturus evolved during Late Miocene times, about six million years ago. The species *Zygomaturus trilobus* is only known from the Pleistocene. During this time it co-existed with *Diprotodon*. These two forms represent the most recently living members of two entirely different diprotodontid lineages, which separated at least 15 million years ago.

Although in overall shape *Zygomaturus* more closely resembled *Diprotodon* than any other contemporary species, it differed in a number of important features. Firstly, it was somewhat smaller. Complete specimens found in Tasmania indicate that *Zygomaturus* was about the same size and shape as a medium-sized hippopotamus, weighing about one tonne. It was nearly two metres long and stood a little over one metre high at the shoulder. The front and back legs, although bearing strong muscles, were quite long. The front feet were particularly broad, and each of the five fingers carried a long, flattened claw. In contrast, the back feet were narrower and weakly clawed, apart from the outermost toe. Like *Diprotodon*, the hind feet of *Zygomaturus* were turned inward and the animal walked on the sides of its feet, probably with a reasonably slow, awkward gait.

The skull of *Zygomaturus* differs from that of *Diprotodon* in a number of significant ways. Although, like *Diprotodon*, the weight of the skull was reduced by the presence of air sinuses, these structures developed in a different way from those of *Diprotodon*, indicating the separate, parallel evolution of the two forms. Although the cheek teeth of *Zygomaturus* indicate that it too was a browser, the teeth were fundamentally different from those of *Diprotodon*. Whereas other diprotodontids possessed relatively simple upper premolars, they were structurally much more complicated in *Zygomaturus*. The difference in tooth structure between *Zygomaturus* and *Diprotodon* was related to different diets and methods of feeding. Unlike *Diprotodon*, the molar teeth of *Zygomaturus* were lower and had broader crests or lophs. The short-faced, powerful-jawed *Zygomaturus* probably ate coarse woody vegetation, such as stems, petioles, roots and bark. Its broad, low-crowned teeth would have functioned in grinding and crushing this type of tough material to a pulp.

Another peculiar feature of *Zygomaturus* was

factf(ocu)s

Zygomaturus was a smaller diprotodontid than *Diprotodon*, being the size of a bullock, nearly two metres long and standing a little over one metre tall. From the distribution of its fossilised remains it seems to have inhabited the colder, wetter parts of the state.

its possession of pre-nasal bones, which may have supported small, horn-like protuberances. Furthermore, a remarkable brow-like flange of bone at the front of the skull is thought to have served as the attachment site for strong muscles that controlled the movement of the upper lip. Quite possibly *Zygomaturus* had snout mobility like that of a pig. Combined with the strong claws on its front feet, the snout may have helped in uprooting whole plants while the *Zygomaturus* grubbed for roots and tubers. Whether or not the horn-like structures assisted in this process in any way is not known. The strong claws on the front feet would have first dug the ground around the plant. The powerful lips would have cleared soil away from the roots. The prominent tusk-like incisors could have firmly grasped the plant, which, helped by the animal's undoubted strength, would then have been ripped from the earth and ground up by the large molar teeth. The occurrence of *Zygomaturus* in the wetter coastal regions of the state — the more richly vegetated areas — may indicate that *Zygomaturus* and *Diprotodon* fed on quite different types of plants.

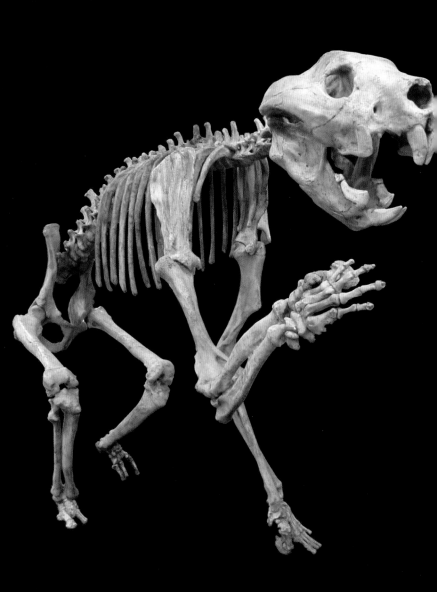

Thylacoleonids

Thylacoleo

As early as the 1830s, a peculiar long, ridged tooth was found in Wellington Caves, New South Wales. This tooth was perplexing to anatomists of the time; however, with the discovery of pieces of skull of the same animal in Victoria in the 1840s it was realised that Australia's fauna had once included a large cat-like marsupial carnivore. It was not until the 1950s that additional bones from this animal were discovered, firstly in South Australia and then in New South Wales, when a nearly complete skeleton was finally unearthed. In Western Australia remains of *Thylacoleo* have been found in the Balladonia region, in Madura Cave on the Roe Plain, in caves on the Nullarbor, and in Mammoth Cave and Devil's Lair. Until 2002 only a few isolated teeth and bones had been found, but an amazing find was made in caves beneath the Nullarbor Plain, when eight complete,

Left: Reconstructed skeleton of *Thylacoleo* (*photo J. Long*)

or nearly complete, skeletons of *Thylacoleo* were discovered, along with extinct kangaroos, thylacines and wombats. All the Western Australian specimens of *Thylacoleo* are a little smaller than those found in the east, and may possibly represent a distinct species. The eastern species is called *Thylacoleo carnifex*.

In terms of size and probable mode of life, *Thylacoleo* should more appropriately be called the marsupial leopard, as its body reached no more than 1.5 metres in length. Its head was relatively large, with a short, wide, heavily built skull. Its short, broad snout had wide, flaring, bony arches on either side. The overall cat-like shape of the head was completed by its forward-looking eyes. Powerful muscles ran from the back of the skull to the lower jaw. The neck was thick and muscular.

Much of our interpretation of what *Thylacoleo* fed upon is gleaned from the nature of its teeth. The most distinctive teeth of *Thylacoleo* were the very large, guillotine-like premolar or carnassial teeth. Each of these teeth may have been nearly six centimetres long, with a sharply bevelled, slightly curved edge. The upper and lower premolar teeth slid against each other in a shearing motion, like a pair of scissors. These teeth lay just below the strengthened cheek buttresses, suggesting that the maximum force of biting centred on this part of the jaw. The jaw also possessed two further premolars, but these were extremely small.

Above: Reconstruction of *Thylacoleo* by Jill Ruse.

The molar teeth were very much reduced in *Thylacoleo*, both in size and number. There was only one in each side of the upper jaw and two in the lower jaw. Their small size argues against them being effective in grinding. The incisors were well developed. The lower ones were bayonet-like and forwardly directed. There were three pairs of upper incisors, the first being very large. These incisors would have formed a most effective set of stabbing pincers. The lower one bore slightly serrated edges. The canine teeth were very small and only a single pair was present in each jaw; this fact was used by some scientists to argue against *Thylacoleo* having had a carnivorous habit. However, as with lions, the jaws were connected in such a way that they could only move up and down; there was no sideways movement and so grinding was not possible.

… Australia's fauna had once included a large cat-like marsupial carnivore.

Thylacoleo had a relatively heavy body, probably weighing up to 150 kilograms, and a strong, but

Right: Skull of *Thylacoleo* from cave in the Nullarbor Plain; one-third natural size *(photo G. Deacon)*.

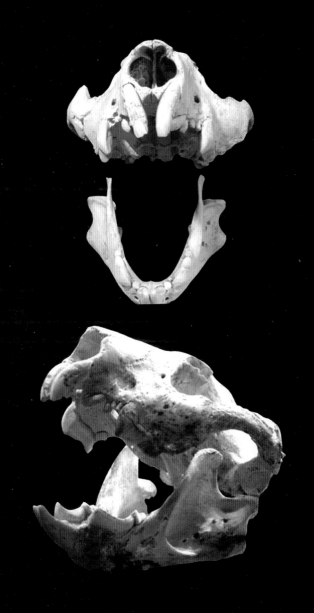

flexible vertebral column. The forefeet were narrow but quite long, with the individual digits parallel to one another and bearing long claws. The 'thumb' was very long and possessed an enormous hooded, curved claw, 3–4 centimetres in length. The paws were heavy and strong. The fore and hind limbs were of similar size. By comparison with living animals, this suggests either a tree-dwelling or running habit, rather than jumping or burrowing. For its body size, the fore limbs were very long and powerful. The large fore paws, with their ferocious claws, were probably used for slashing and tearing at their prey. The hind feet possessed semi-opposable toes, which would have helped in climbing trees. The strong limbs suggest adaptation for moderately active running and prey catching.

Richard Owen, who first described *Thylacoleo* in 1859, was firmly of the opinion that it was a carnivore. Although it lacked effective canine teeth, which are generally found in carnivores, it also lacked effective grinding molars, a prerequisite for herbivores. Since Owen's first pronouncements, *Thylacoleo* has at various times been interpreted as either a carnivore, a herbivore, or a scavenging 'bone crusher', although little evidence has been presented for this last interpretation.

Those who favour a herbivore interpretation have offered various suggestions as to *Thylacoleo*'s diet: anything from soft fruit to cycad nuts. It has even been suggested that they fed on cucumbers or crocodile eggs. Others have thought that *Thylacoleo* might have eaten twigs, fine branches or leaves. However, the lack of effective molars makes this unlikely; even if the premolars could have cut off the vegetation, the teeth would have been incapable of chewing the course, fibrous material.

The general view held at present is that, like

factfocus

For 150 years scientists have debated the functional significance of the strange teeth of *Thylacoleo*, especially its huge, slicing premolars. Many have thought these indicate *Thylacoleo* had a carnivorous diet, but others have suggested it fed on soft fruit, or cycad nuts, crocodile eggs, or even branches or leaves. On balance it seems most likely that it was a carnivore.

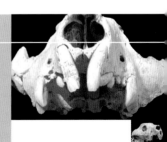

lions and cheetahs, *Thylacoleo* may have killed by strangling or suffocating its prey, using its large incisors. Stabbing with these incisors may also have caused the prey to bleed to death. The large claw on the heavy paws, which was attached to the strong forelimb, would have allowed *Thylacoleo* to knock its prey to the ground and hold it down while strangling it. The large premolars would have been used to slice through the flesh, while the incisors would have pulled the flesh from the carcass. If arboreal (tree-dwelling) and carnivorous, *Thylacoleo* is likely to have fed on some of the larger-sized herbivores, perhaps pouncing on them from trees. It probably also fed on smaller herbivores such as possums and koalas.

Thylacoleo may have killed by strangling or suffocating its prey …

Little is known of the origins of *Thylacoleo*. It probably evolved from an earlier form, called *Wakaleo*, of Late Miocene age (approximately 10 million years). Of living marsupials, *Thylacoleo* is most closely related to the wombat.

Above: Skull and bones of *Thylacoleo* as they were found in a cave in the Nullarbor Plain *(photo C. Bryce)*.

Wombats

The only place where wombats occur in Western Australia today is close to the border with South Australia, in the eastern part of the Nullarbor. This is the western-most extension of the range of the hairy-nosed wombat, *Lasiorhinus latifrons*, although cave deposits indicate that it once extended across much of the Nullarbor, until perhaps very recently.

During the Pleistocene, however, extinct species of wombats roamed over much of the state. Remains of a very large wombat, called *Phascolonus gigas*, have been found as far apart as the Kimberley and the south of the state at Balladonia. Another species, *Vombatus hacketti*, which was very similar in size and form to the other living species of wombat, *Vombatus ursinus*, inhabited south-western Australia during the latter part of the Pleistocene.

Phascolonus gigas was the largest wombat and

Left: Modern wombat (photo *©iStockphoto.com/5cheherazade*).

possibly the largest burrowing marsupial ever to have lived. Its body size was at least twice that of any modern wombat, being about 1.7 metres in length. It would have stood about one metre high at the shoulder and weighed up to 200 kilograms. In addition to its overall size, *Phascolonus gigas* was also characterised by its relatively large head. At 40 centimetres in length, it was similar in size to a cow. The limbs, although short, were stout and strong. The feet are poorly known. The distal end of the humerus of *Phascolonus gigas* was relatively wider than in living wombats, suggesting that this massive wombat was equally adept at burrowing as the modern-day forms.

> A typical modern-day wombat eats roots, shoots and leaves.

Phascolonus gigas differed from all other wombats in the nature of its teeth. Like other wombats the molar teeth were bi-lobed, open-rooted and curved; however, the upper incisors were also long and

Right: Reconstruction of *Phascolonus gigas* by Jill Ruse.

curved, as well as being very wide. They would seem to fit into the mouth of a *Diprotodon* more readily than that of a wombat. The general diet of *Phascolonus gigas* probably comprised mainly grass and leaves. A typical modern-day wombat eats roots, shoots and leaves.

Phascolonus gigas was widespread across much

of Australia during the Pleistocene, but it seems to have been adapted to semi-arid conditions. Fossil remains have been found in the north-west of the state, as well as in caves in the south-west and Nullarbor Plain.

Wombats are evolutionarily closest to the extinct diprotodontids. Amongst living marsupials their closest relative is the koala. Fossilised remains of extinct wombats, known from rocks elsewhere in Australia, are older than 10 million years.

Right: Skull and lower jaw of *Vombatus hacketti* from Mammoth Cave; two-thirds natural size *(photo G. Deacon)*.

factf⊙*cus*

Although today wombats are only found in Western Australia close to the South Australia border, they were widespread during the Pleistocene. Two species occurred: *Vombatus hacketti*, a species similar to the living wombat, and *Phascolonus gigas* which, as its name implies, was a giant among wombats, weighing about 200 kg.

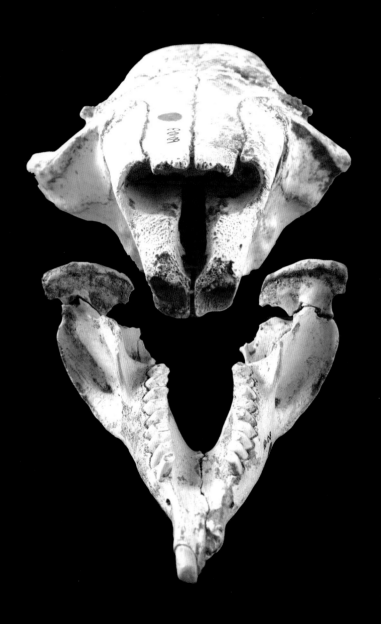

Thylacines

The thylacine (also known as the Tasmanian tiger, or wolf), *Thylacinus cynocephalus*, which existed in Tasmania until 1936, once occurred commonly on the mainland of Australia and in Papua New Guinea. In Western Australia its remains have been found at Tunnel Creek in the Kimberley, the Cape Range, in caves between Moore River and Dongara, Yanchep, Mammoth Cave, Strong's Cave, Devil's Lair and numerous caves on the Nullarbor.

The thylacine was a large carnivorous marsupial, similar in appearance to a dingo but with relatively shorter legs, sandy-yellow hair, with 15–20 prominent dark-brown stripes up to three centimetres wide across the shoulders, back and tail. The tail was long and stiff. Its hind legs looked very crooked when compared with those of a dog.

The Tasmanian form of the thylacine averaged

Left: Thylacines at Hobart Zoo circa 1920s. (*Top: Photographer unknown. Bottom: B. Sheppard.*)

about 25 kilograms in weight and had a head and body length of a little over one metre. The tail was about 0.5 metres in length. Although the thylacine normally walked on four legs, it was able to rise up on its hind legs. Its head was quite wolf-like, but it had shorter ears than a wolf. Its jaws were long and powerful, and could open very wide. The female thylacine produced litters of up to four young, which were dependent on the mother until they were about half grown. The prey of the thylacine was mainly kangaroos, possums, bandicoots, small rodents and birds, which they generally hunted at night.

Above: Similarity due to convergent evolution between the thylacine (top) and dingo (below); two-thirds natural size *(photos D. Elford)*.

Left: Reconstruction of *Thylacinus cynocephalus* by Jill Ruse.

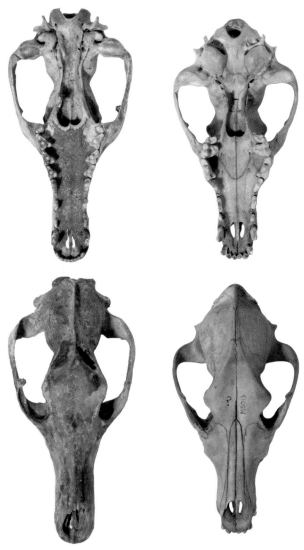

Above: Similarity between the skulls of the thylacine (left) and dingo
(right); one-third natural size *(photos D. Elford)*.

Comparison of skull material of the Western Australian form with Tasmanian specimens shows that the Western Australian form was a little smaller. In addition to bone material from the cave deposits, some specimens have been found with skin and hair still present. Radiocarbon dating of this material has shown that the thylacine was alive in Western Australia until about 3,100 years ago. One specimen found in Thylacine Hole, a cave on the Nullarbor, is an almost intact mummified specimen, which has been dated at about 4,500 years. In addition to retaining virtually all of its skin and hair, this specimen even has its eyes and tongue preserved. Mummified carcasses of other animals have also been found in this cave. Compared with other caves Thylacine Hole

factfocus

When a complete mummified carcase of a thylacine was found in a cave on the Nullarbor in the 1960s it was thought that this was evidence that thylacines were still living on the mainland of Australia. Not only was it covered in fur, but its eyeballs, tongue and whiskers were all intact. However, radiocarbon dating found it had died about 4,500 years ago.

has a high salt content, suggesting that this was the reason for the thylacine's natural mummification and lack of decay. Thylacine Hole has a shaft 1–2 metres in diameter that drops vertically for about 12 metres to the cave floor. The thylacine was found 150 metres from the bottom of the shaft; thus it must have survived its fall and wandered to the far end

The dingo probably outcompeted the thylacine …

of the cave before presumably dying of starvation. A number of other, near-complete skeletons of thylacines have also been found in this cave.

The co-existence of the thylacine with humans on the mainland of Australia is shown by Aboriginal drawings and paintings of a striped, dog-like animal. The disappearance of the thylacine from the mainland may well have been related to the appearance of the dingo in Australia. The oldest reliable radiocarbon date for the dingo is a little over 3,000 years before present. The dingo probably out-competed the thylacine, perhaps by being a more efficient hunter

Above: 4,500 year old mummified thylacine from a cave on the Nullarbor Plain; specimen on display in the Western Australian Museum *(photo A. Baynes).*

and also breeding more rapidly. The persistence of the thylacine in Tasmania into historical times supports this idea because the dingo never reached Tasmania. The dingo and thylacine are known to have co-existed on the mainland, bones of the two species having been found on the same piece of rock in North West Cape.

What the dingo probably achieved on the mainland of Australia, European settlers achieved in Tasmania in not much more than 100 years. In the 31 years between 1878 and 1909, for instance, a total of 4,821 thylacines were killed in Tasmania, the bounty on each adult being one pound. It is a trifle ironic to note that in 1936, just under two

factf⊚cus

Three thousand years ago the dingo arrived in Australia from south-east Asia. About this time the last thylacine became extinct on the Australian mainland. The dingo never colonised Tasmania, and the last thylacine died there, in Hobart Zoo, in 1936.

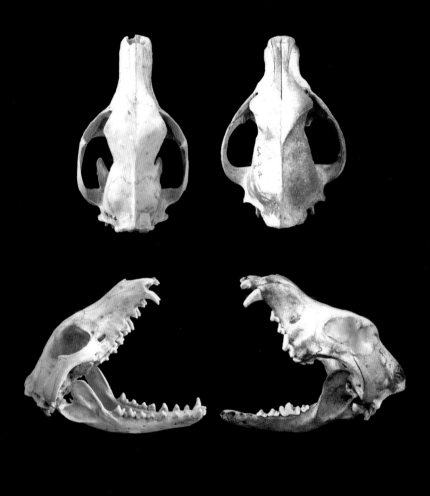

Above: The close similarity between the thylacine (left) and the dingo (right), two completely unrelated mammals; one-quarter natural size (*photo G. Deacon*).

Above: Mummified thylacine from a cave on the Nullarbor Plain;
specimen on display in the Western Australian Museum
(photo A. Baynes).

months after the thylacine was finally added to the list of wholly protected animals in Tasmania, the last known individual died, on 7 September 1936. With its passing, a whole family of marsupials that had existed for more than 24 million years became extinct.

The thylacine was remarkably similar in appearance to a group of extinct South American marsupials called the borhyaenids, and scientists once suggested that the two groups were directly related. However, this is an example of convergent evolution, because it has been shown that the thylacine was closely related to the Australian dasyurids.

Above: Entering cave on the Nullarbor Plain through a sinkhole
(photo C. Bryce).

Following Page: Carefully collecting the remains of extinct
marsupials in a cave on the Nullarbor Plain (photo C. Bryce).

The half a million year resting place
of a menagerie of extinct mammals.

Kangaroos

During the Pleistocene, the kangaroo, or macropodid, fauna of Australia was more diverse than it is at present, being represented, in particular, by a greater number of large forms. Many of these far exceeded in size the largest of the living kangaroos. Macropodids can be divided into two major groups: the macropodines, including *Macropus*, *Wallabia* and *Protemnodon*; and the sthenurines, including *Simosthenurus* and *Procoptodon*. In Western Australia the most diverse fauna occurs in the Early Pleistocene cave deposits in the Nullarbor, where 23 species of kangaroos have been recovered.

Macropus

All of the large kangaroos in Australia today belong within the genus *Macropus*. At present there are 14 living species of *Macropus*, and a similar number of extinct forms is known. The earliest known kangaroo

Left: Modern red kangaroo (*photo ©iStockphoto.com/kcphotoman*)

occurs in fossil deposits about eight million years old and had teeth similar in size and shape to the quokka, *Setonix*. About 30 fossil and living species have evolved within the last five million years. Like living species, the fossil species were all grazing animals, and the appearance of so many species in a relatively short period of time, geologically speaking, is thought to have occurred in response to a great increase in the extent of grasslands in Australia as the climate became progressively drier. A modification in the structure of the teeth of these *Macropus* species allowed them to cope efficiently with the tough grasses which dominated the grassland areas.

The fossilised remains of two large extinct

The Pleistocene mammal fauna was characterised by the presence of kangaroos far bigger than today's. Some species ancestral to the eastern grey kangaroo stood about 2.5 metres tall, while the extinct short-faced kangaroo *Procoptodon* reached about 3 metres in height.

species of *Macropus* have been found in Western Australia. *Macropus pan*, a Late Pliocene form, 2–2.5 million years old, has been found at Quanbun in the Kimberley. This species resembles the living wallaroo, *M. robustus* and the antilopine kangaroo, *M. antilopinus*, a northern Australian species. However, *M. pan* was much larger, reaching a height in excess of two metres.

Other living kangaroos are thought to have had even larger ancestors. The eastern grey kangaroo, *M. giganteus*, an eastern states species, is thought to have evolved from *M. titan*, a Late Pleistocene species of kangaroo that was about 30 percent larger than *M. giganteus* and reached a height of about 2.5 metres.

Remains of a large species of kangaroo have been found in Western Australia in the deposits at Balladonia and Mammoth Cave. It is possible that this form, similar in size to *M. titan*, may have been the ancestor of the western grey kangaroo, *M. fuliginosus*. However, at present it is not known for certain to what species the remains should be assigned.

Protemnodon

In cave deposits in the south-west (such as Mammoth Cave, Crystal Cave, Strong's Cave, Hastings Cave) and at Leighton Beach, Gingin, Quanbun and in some caves in the Nullarbor, the fossilised remains of an extinct genus of kangaroo called *Protemnodon* have been found. Most of the 12 known species were as large as or larger than the living grey kangaroo and were characterised by their heavy

Above: Part of the lower jaw of the extinct giant wallaby *Protemnodon* (left) compared with the equivalent part of the jaw of a modern wallaby (right); natural size *(photo D. Elford).*

build; short, stout feet; and relatively flat, elongate skulls, which lack the characteristic downward curve of the snout found in modern kangaroos. The shape of the feet suggests that locomotion must have been quite slow. *Protemnodon anak*, the smallest of the three species whose remains have been found in Western Australia, reached about 1.5 metres standing height and weighed about 40 kilograms. It had a relatively long neck and large

Left: Reconstruction of *Protemnodon* by Jill Ruse.

powerful forearms. Stomach contents have been found in some specimens. These consist of coarse twigs and large fragments of vegetation, indicating that this species of *Protemnodon* (and possibly all the others) was a browser. It would have used its powerful forearms to pull branches down to a height at which it could comfortably feed. Some of the teeth of this

> … sthenurines differ from the macropodines in the structure of the foot.

species were well adapted for shearing through stems and petioles. The tooth structure suggests that this particular species of *Protemnodon* may also have been capable of some grazing.

In Mammoth Cave, the remains of a larger species of *Protemnodon* have been found. This species, *P. brehus*, was similar in form but a little smaller than the largest of the *Protemnodon* species to have been found in Western Australia, a form from Quanbun which is thought to be *P. roechus*. This species was probably comparable in mass to *Macropus titan*.

Right: Reconstruction of *Simosthenurus* by Jill Ruse.

Simosthenurus and *Procoptodon*

During the Pleistocene in Western Australia, large sthenurines belonging within *Simosthenurus* and *Procoptodon* (the 'short-faced' kangaroos) were relatively common animals, their remains having been found in most fossil deposits yielding Pleistocene mammal remains. The sthenurines differ from the macropodines in the structure of the foot. In sthenurines, all the digits of the hind foot were reduced in size, except for the fourth digit. In some

sthenurines the reduction was greater even than that of the modern horse, the claw of the fourth digit of *Simosthenurus* being almost hoof-like in appearance. In macropodines the digits are less reduced, and the fifth is functionally important and relatively large. The two groups also differ in the nature of their teeth. In the macropodines the enamel on the teeth is smooth, whereas in sthenurines it is crenulated.

The skull of *Simosthenurus* is shorter than in all other kangaroos, apart from the giant *Procoptodon*. Its eyes are more forwardly pointed than in other kangaroos. *Simosthenurus* was probably a browser, its crenulate, ornamented molars and structurally complex premolars being well suited to chewing leaves and stems.

factfocus

One group of kangaroos that now no longer exists is the short-faced kangaroos or sthenurines. Unlike modern kangaroos, these were predominantly browsers, not grazers. As well as the short face they are characterised by the presence of a large, hoof-like toe, the other toes having reduced greatly in size.

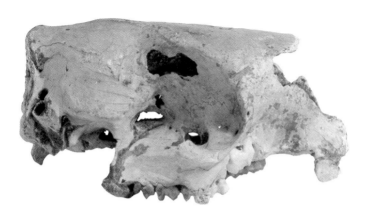

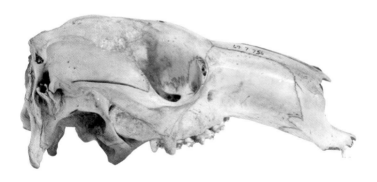

Above: Profiles of skulls of *Simosthenurus brownei* (top) and the living western grey kangaroo, *Macropus fuliginosus* (bottom); half natural size *(photo D. Elford).*

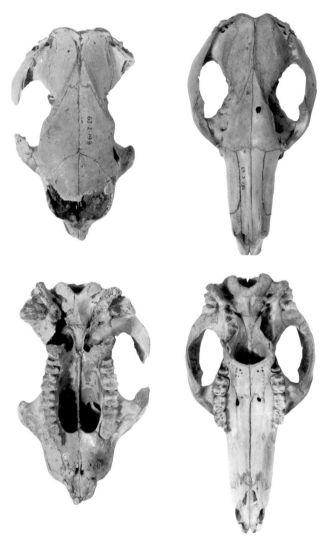

Above: Skulls of *Simosthenurus brownei* (left) and *Macropus fuliginosus* (right); one-third natural size *(photo D. Elford)*.

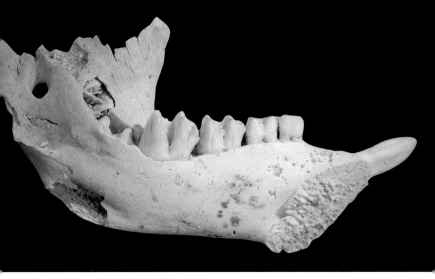

Above: Lower jaw of the giant sthenurine kangaroo *Procoptodon goliah*; one-half natural size *(photo G. Deacon)*.

The largest of all kangaroos was the sthenurine, *Procoptodon goliah*. This species probably stood up to three metres tall and weighed up to 250 kilograms. It is thought to have used its long, powerful arms to pull down branches; its teeth were adapted for eating coarse vegetation. Its very deep, short-faced skull had forwardly directed eyes and, like other sthenurines, its feet were horse-like, with a greatly enlarged fourth toe. A single specimen of this species found in a cave on the Nullarbor attests to the presence of this giant kangaroo in Western Australia during the Pleistocene.

In Western Australia two true species of

Simosthenurus are known from fossil deposits: *S. occidentalis* and *S. maddocki*. A species formerly assigned to *Simosthenurus*, *S. browneorum*, appears to be intermediate between *Simosthenurus* and *Procoptodon* and has recently been referred to as '*Procoptodon*' *browneorum*. Another of these intermediate forms found in a Nullarbor cave is '*Procoptodon*' *williamsi*. Two other species, which, while intermediate, are closer to *Simosthenurus* are '*S.*' *pales* and '*S.*' *baileyi*. In terms of size, species such as *S. occidentalis* and '*P.*' *browneorum* were similar in height but much heavier than the western grey kangaroo, *Macropus fuliginosus*.

Species of *Simosthenurus* and *Procoptodon*, and inter-mediate forms, have been found in Western Australia in caves in the south-west, such as Mammoth Cave, Crystal Cave and Strong's Cave; at Yanchep, Wanneroo, Gingin and Balladonia; and in caves in the Nullarbor.

Borungaboodie

This recently described potoroid kangaroo is the largest of the small wallabies known as bettongs.

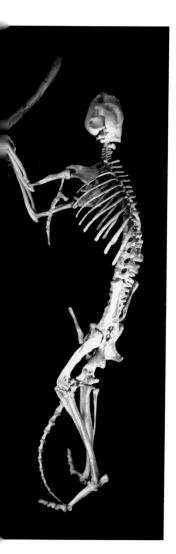

It is thought to have weighed up to 10 kilograms, which is 30 percent more than the largest living bettong, the rufous bettong, *Aepyprymnus rufescens*. Although only known from a single jaw, *Borungaboodie hatcheri* is markedly different from living bettongs. In addition to being much larger, it has a number of unique dental features. This has led to speculation that while it probably fed on a diet of nuts or stone fruit, it may also have supplemented its diet with meat when the opportunity arose. It might have scavenged on meat or killed small vertebrates as modern bettongs are known to do.

Left: Reconstruction of *Simosthenurus* browsing. In 'Diamonds to Dinosaurs' gallery, W.A. Museum *(photo G. Deacon)*.

Echidnas

The rich bone deposits in Mammoth Cave have yielded the bones of three species of echidna. These deposits probably contain the most diverse monotreme (egg-laying mammal) fauna found at any one locality in Australia. In addition to the remains of the living Australian species of echidna, *Tachyglossus aculeatus*, some limb bones of a very large species of echidna have been found in Mammoth Cave. This form, known as *Zaglossus hacketti*, was the largest known monotreme, which was about one metre in length, stood 0.5 metres at the shoulder and weighed about 30 kilograms. The third species, *Megalibgwilia ramsayi*, was intermediate in size between the other two species, having been about 0.6 metres in length and weighing about 10 kilograms. This echidna was very similar to the only other living species, *Zaglossus bruijni*, which lives in the highland regions of Papua New Guinea.

Left: Modern echidna (*photo ©iStockphoto.com/clearviewstock*)

Zaglossus differs from *Tachyglossus* in the shape of its tongue and skull as well as its much larger body size. The snout is relatively larger in *Zaglossus* than *Tachyglossus*, and the animal has a more erect stance. Unlike *Tachyglossus*, which feeds mainly on ants and termites, the diet of *Zaglossus* is largely restricted to earthworms. *Megalibgwilia ramsayi* differed from the living species of *Zaglossus*, principally in its relatively sturdier, shorter, stouter and straighter snout and longer forearms. This may mean that it fed on something other than earthworms. The shorter, broader, less down-curved back; and compact long bones with more prominent flanges and lever arms for muscle attachment would have allowed *Megalibgwilia ramsayi* to pry open large logs and termite nests. Its feeding behaviour may have been like that of *Tachyglossus* but on a grander scale.

The very large echidna, *Zaglossus hacketti*, should probably be placed in a separate genus from either *Zaglossus* or *Tachyglossus*, not only because of its

Above: Reconstructions of *Zaglossus hacketti* (top), *Megalibgwilia ramsayi* (bottom left hand) and *Tachyglossus aculeatus* (bottom right) by Jill Ruse.

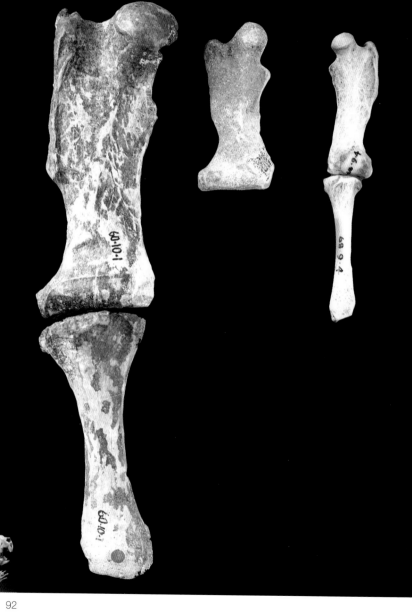

much larger size (its limb bones being nearly three times the length of those of *Tachyglossus*) but also because of its different limb proportions. However, neither its skull nor jaws have ever been found. The limb proportions suggest it walked in a different way to the other echidnas; it possessed a relatively long femur, which resulted in the centre of gravity of the body being set further back than in the smaller species of echidna. This may have allowed more freedom for the forelimbs to dig, and may even have permitted the animal to stand on its hind legs while feeding at ant or termite mounds.

Although the large forms of echidna became extinct in Australia, one species, *Zaglossus bruijni*, has persisted in Papua New Guinea. This species has managed to survive intense hunting pressure for thousands of years, which suggests that humans were probably not directly responsible for the disappearance from Australia of *Zaglossus* or *Megalibgwilia*. The change in the relative proportion of grasslands and forests in Australia during the Late

Left: Leg bones of *Zaglossus hacketti* (left), *Megalibgwilia ramsayi* (centre) and *Tachyglossus aculeatus* (right); three-quarters natural size *(photo G. Deacon)*.

Pleistocene, due to the effects of polar glaciation, restricted the range of closely related species that could coexist with one another. Competition would have intensified, resulting in the extinction of larger species, which were presumably less adaptable to these changes in climate and vegetation. *Zaglossus* survived in highland Papua New Guinea, possibly by migrating up and down mountains, shifting with the changing zones of vegetation.

Right: John Long (left) and Carmelo Amalfi (right) applying glue to harden the bones of *Thylacoleo*. (*photo C.Bryce*)

factfocus

The living echidna pales into insignificance when compared with the extinct Pleistocene species *Zaglossus hacketti*. This giant among monotremes was about one metre long, standing half a metre at the shoulder, and weighing in at about 30 kg.

Mammal Extinctions in Western Australia

There have been three principal phases of extinctions of mammals in south-western Australia during the last 100,000 years. The first of these saw the disappearance of many of the larger forms, the so-called 'megafauna'. The timing of this extinction event has been the subject of much debate. Recent studies indicate that the megafauna had become extinct throughout Australia between 40–50,000 years ago.

The animals which became extinct in this first phase of extinctions in Western Australia include most of the large-bodied forms, such as *Zygomaturus*, *Diprotodon*, *Protemnodon*, *Simosthenurus*, *Procoptodon*, *Thylacoleo*, *Zaglossus* and large forms of *Macropus*. Furthermore, two forms still living in eastern Australia — the koala, *Phascolarctos*, and the woodland wombat, *Vombatus* — became extinct in Western Australia at this time. Many small mammal

Left: (*photo C.Bryce*)

species still living, at least in historical times, are known from the Mammoth Cave deposits and presumably co-existed with the larger, now extinct, mammals. These include the quoll (*Dasyurus*), wambenger (Phascogale), marsupial 'mouse' (Antechinus), dunnart (Sminthopsis), the short-nosed bandicoot (*Isoodon*), the quokka (*Setonix*) and some kangaroos and wallabies, including the tamar wallaby (*Macropus eugenii*), the western brush wallaby (*Macropus irma*) and the western grey kangaroo (*Macropus fuliginosus*).

The second phase of extinction in the extreme south-west of Australia saw the disappearance of forms today occurring in other parts of the continent, apart from the thylacine, which probably became extinct in historical times. The forms that disappeared from the south-west in this second phase of extinction went within the last 15,000 years. They include, in addition to the thylacine, the Tasmanian devil (*Sarcophilus*), the long-nosed bandicoot (*Perameles*), the boodie (*Bettongia lesueur*), the rock wallaby (*Petrogale*), the hare-wallaby (*Lagorchestes*), some native mice (*Pseudomys occidentalis* and

Above: Lower jaw of a koala, *Phascolarctos cinereus* from Mammoth Cave *(photo D. Elford).*

Pseudomys albocinereus), hopping mice (*Notomys*) and the ghost bat (*Macroderma*).

The third phase of extinction began a little over two hundred years ago and is continuing today. When Europeans arrived in Australia they brought with them cats and foxes. These, along with habitat destruction caused by clearing of the vegetation, primarily for agriculture, have had a catastrophic effect on the small mammal fauna that had managed to survive the two previous extinction episodes. Studies of fossil bone deposits from owl roosts suggest that perhaps three-quarters of Australia's small mammal species have gone extinct in the last 200 years.

Although the reasons for this most recent phase of extinction are clear, the disappearance of the large mammals in the first extinction phase has been the subject of much debate. Two principal agents have been blamed: climate change or human activity. There is little doubt that during the last 100,000 years the Earth has suffered major climatic changes associated with the last great Ice Age, which peaked some 15–12,000 years ago. As the ice caps advanced from about 80–25,000 years ago, the Earth became progressively drier as water became locked in the ice caps. It has been argued that evolution of mammals of large body size during the Cenozoic occurred as an adaptation to poorer food as aridity increased with the onset of the Pleistocene glaciations.

Such a trend toward increasing body size also occurred at this time in other parts of the world. During the most arid period, from 25–10,000 years before present, the largest species, it has been argued, would have been under the greatest stress and thus were more likely to have suffered extinction than smaller forms. A relatively rapid increase in aridity during the Late Pleistocene would have changed the environment to such an extent that many species would have been unable to adapt to the new conditions. These environmental changes affected both the climate and vegetation. The cooler, drier conditions of the Late Pleistocene would have resulted in substantial changes to the type and distribution of vegetation, in particular the changing proportions of grassland and forests. However, in Western Australia as well as in the rest of Australia the extinction of the megafauna had occurred before the most intense arid period 20–15,000 years ago. Significantly, the megafauna had survived similar arid conditions during earlier glacial periods, so what was so different about this last glacial period?

factfocus

So what caused the extinction of the Australian Pleistocene megafauna? We don't know for certain. Main contenders are climate change, with the onset of extreme aridity 40 – 50,000 years ago; the influence of the first humans to arrive in Australia at this time, either through hunting, or changing the mammals habitats; or a combination of both factors.

During this period of environmental upheaval a new and very effective predator arrived in Australia from southeast Asia — humans. They are thought to have first arrived in Australia between 50–60,000 years ago. Their affect on the environment is likely to have been profound, as a result of both their predatory activity and their techniques of frequently burning areas of vegetation as a means of hunting animals. Such extensive burning would have had the effect of radically altering the vegetation composition and perhaps exacerbating the decline of forests, which was already occurring as a result of the increased aridity. It is possible that humans merely accelerated an inevitable event; however, it is clear that on the basis of recent evidence the megafauna became extinct within less than 10,000 years of the arrival of humans. Recent evidence suggests that the megafauna persisted on Tasmania for longer than on the mainland, specimens of *Protemnodon anak* having been dated at about 41–43,000 years old.

It is possible that humans merely accelerated an inevitable event.

Significantly, humans did not arrive on the island until about 43,000 years ago. Following their arrival the megafauna seems to have disappeared, probably due to direct hunting.

Proving whether or not humans actually co-existed with the extinct megafauna or not has been difficult to demonstrate. However, in Mammoth Cave bones of *Simosthenurus* have been identified which are notched and charred, or broken in such a way to indicate deliberate breakage over a hard edge. These bones are interpreted as having been in some way tampered with by the early inhabitants of Australia. They are one of the few cases of direct evidence for interaction between humans and the extinct megafauna.

Maybe neither humans alone nor the increasing aridity were solely responsible for the mammal extinctions, but both factors might have caused the loss of the diverse mammal fauna. The added impact of humans on top of an already stressed environment may well have been the final straw that broke the camel's (or *Diprotodon*'s) back.

Acknowledgements

I wish to thank Dr DJ Kitchener, Dr A Baynes, Dr T Flannery, Dr T Rich, Mr GW Kendrick, Mrs NO McNamara, Ms S Elliott and Dr G Prideaux for reading various versions of this manuscript. Dr A Baynes, Mr D Elford, Ms S Elliott, Dr G Deacon, Dr J Long and Mr C Bryce are thanked for the photographs. I am very grateful to Cathie Glassby for her excellent design of the book and to Jill Ruse for the beautiful reconstructions which, although original, were made easier by reference to Mr P Schouten's excellent drawings in *Prehistoric Animals of Australia*.

Further Reading

Archer, M. (1972). '*Phascolarctos* (Marsupalia, Vombatoidea) and an associated fossil found from Koala Cave near Yanchep, Western Australia.' *Helictite* 10: 45–59.

Archer, M. and Clayton, G. (eds) (1984). *Vertebrate Zoogeography and Evolution in Australia*. Hesperion Press, Perth.

Dortch, C. (1984). '*Devil's Lair: A Study in Prehistory*.' WA Museum, Perth.

Finch, M.E. (1971). '*Thylacoleo* marsupial lion or marsupial sloth?' *Australian Natural History* 17: 7–11.

Finch, M.E. (1982). 'The discovery and interpretation of *Thylacoleo carnifex* (Thylacoleonidae, Marsupalia).' Pp. 537–551 in *Carnivorous Marsupials*, (ed) M. Archer. Royal Zoological Society of NSW.

Finch, M.E. and Freedman, L. (1982). 'An odontometric study of the species of *Thylacoleo* (Thylacoleonidae, Marsupalia).' Pp. 553–561 in *Carnivorous Marsupials*, (ed) M. Archer. Royal Zoological Society of NSW.

Flannery, T.F. (1984). 'Re-examination of the Quanbun Local Fauna, a Late Cenozoic vertebrate fauna from Western Australia.' *Records of the Western Australian Museum* 11: 119–128.

Glauert, L. (1948). 'The cave fossils of the south-west.' *Western Australian Naturalist* 6: 100–104.

Long, J., Archer, M, Flannery, T. and Hand, S. (2002). *Prehistoric Mammals of Australia and New Guinea*. University of New South Wales Press, Sydney.

Lowry, D. C. and Lowry, J.W.J. (1967). 'Discovery of a Thylacine (Tasmanian Tiger) carcase in a cave near Eucla, Western Australia.' *Helictite* 5: 25–29.

Lowry, J.W.J. and Merrilees, D. (1969). 'Age of the desiccated carcase of a Thylacine (Marsupalia, Dasyuroidea) from Thylacine Hole, Nullarbor Region, Western Australia.' *Helictite* 7: 15–16.

Merrilees, D. (1967). 'South-western Australian occurrences of *Sthenurus* (Marsupalia, Macropodidae), including *Sthenurus brownei* sp. nov.' *Journal of the Royal Society of Western Australia* 50: 65–79.

Merrilees, D. (1968). 'Man the destroyer: Late Quaternary changes in the Australian marsupial fauna.' *Journal of the Royal Society of Western Australia* 51: 1–24.

Merrilees, D. (1979). 'The prehistoric environment in Western Australia.' *Journal of the Royal Society of Western Australia* 62: 109–128.

Murray, P. (1978). 'Late Cenozoic monotreme anteaters.' Pp. 29–56 in *Monotreme Biology*, (ed) G.B. Sharman. Royal Zoological Society of NSW.

Murray, P. (1978). 'Australian megamammals: Restorations of some Late Pleistocene fossil marsupials and a monotreme.' *The Artefact* 3: 77–99.

Murray, P. (1984). 'Extinction downunder: A bestiary of extinct Australian Late Pleistocene monotremes and marsupials.' In *Pleistocene Extinctions*, (ed) P. Martin and R. Klein. University of Arizona.

Murray, P. (1984). *Australia's Prehistoric Animals*. Methuen, Sydney.

Prideaux, G.J. (1999). *Borungaboodie hatcheri* gen et. sp. nov., a very large bettong (Marsupalia: Macropodoidea) from the Pleistocene of southwestern Australia. *Records of the Western Australian Museum Supplement* 57: 317–329.

Prideaux, G.J. (2004). 'Systematics and evolution of the sthenurine kangaroos.' *University of California Publications in Geological Sciences* 146: v–xvii, 1–622.

Prideaux, G.J., Long, J.A., Ayliffe, L.K., Hellstrom, J.C., Pillans, B., Boles, W.E., Hutchinson, M.N., Roberts, R.G., Cupper,

M.L., Arnold, L.J., Devine, P.D. & Warburton, N.M. (2007). 'An arid-adapted middle Pleistocene vertebrate fauna from south-central Australia.' *Nature* 445: 422–425

Quick, S. and Archer, M. (1983). *Prehistoric Animals of Australia*. Australian Museum, Sydney.

Rich, P.V. and Thompson, E.M. (eds) (1982). *The Fossil Vertebrate Record of Australasia*. Monash University, Melbourne.

Roberts, R.G., Flannery, T.F., Ayliffe, L.K., Yoshida, H., Olley, J.M. Prideaux, G.J., Laslett, G.M., Baynes, A., Smith, M.A., Jones, R. & Smith, B.L. (2001). 'New ages for the last Australian megafauna: Continent-wide extinction about 46,000 years ago.' *Science* 292: 1888–1892.

Roberts, R.G. & Brook, B.W. (2010). 'And then there were none?' *Science* 327: 420–422.

Smith, S. (1981). 'The Tasmanian Tiger — 1980.' National Parks and Wildlife Service of Tasmania, *Technical Report* 8717.

Tedford, R.H. (1973). 'The Diprotodons of Lake Callabona.' *Australian Natural History* 17: 349–354.

Turney, C. S. M., Flannery, T.F., Roberts, R.G., Reid, C., Fifield, L.K., Higham, T.F.G., Jacobs, Z., Kemp, N., Colhuon, E.A. Kalin, R.M. & Ogle, N. (2008). 'Late-surviving megafauna in Tasmania, Australia, implicate human involvement in their extinction.' *Proceedings of the National Academy of Sciences USA* 105: 1210–125.

Webb, S. (2008). Megafauna demography and late Quaternary climatic change in Australia: a predisposition to extinction. *Boreas* 37: 329–345.

Wroe, S., Myers, T. J., Wells, R. T. & Gillespie, A. (1999). Estimating the weight of the Pleistocene marsupial lion, *Thylacoleo carnifex* (Thylacoleonidae : Marsupialia): implications for the ecomorphology of a marsupial super-predator and hypotheses of impoverishment of Australian marsupial carnivore faunas. *Australian Journal of Zoology* 47: 489–498.

First Published 2010 by the
Western Australian Museum
49 Kew Street, Welshpool, Western Australia 6106
(Postal: Locked Bag 49, Welshpool DC. WA 6986)
www.museum.wa.gov.au

Designer Cathie Glassby
Printed by Everbest Printing Company, China.

National Library of Australia
Cataloguing-in-publication entry
Author: McNamara, Ken.
Title: Prehistoric mammals of Western Australia / Ken
McNamara, Peter Murray.
Edition: Rev. ed.
ISBN: 9781920843540 (pbk.)
Series: Factfocus.
Subjects: Mammals, Fossil — Western Australia.
Paleontology — Western Australia.
Other Authors/Contributors: I. Murray, Peter.
II. Western Australian Museum.
Dewey Number: 569.2